Rocks and Soil

How Weather and Other Forces Change the Earth

DEVELOPED IN COOPERATION WITH

CRANBROOK INSTITUTE OF SCIENCE
BLOOMFIELD HILLS, MICHIGAN

Scholastic Inc. grants teachers who have purchased *Science Place* permission to reproduce from this book those pages intended for use in their classrooms. Notice of copyright must appear on all copies of copyrighted materials. Other than for such use, no part of this publication may be reproduced in whole or in part, or stored in a retrieval system, or transmitted in any form or by any means, electronic, mechanical, photocopying, recording, or otherwise, without written permission of the publisher. For information regarding permission, write to Scholastic Inc., Instructional Publishing Group, 555 Broadway, New York, NY 10012.

Copyright © 1997, 1993 by Scholastic Inc. All rights reserved. Published by Scholastic Inc.
Printed in the U.S.A.
ISBN 0-590-95482-2
1 2 3 4 5 6 7 8 9 10 09 03 02 01 00 99 98 97 96

EARTH, WITHIN THE UNIVERSE, IS CONSTANTLY CHANGING.

Rocks and Soil

Earth is changing.

Read-Aloud

Rocks and Soil

Products form as Earth changes.

Certain processes and forces cause Earth to change.

Literature

Literature

Earth changes over long periods of time.

What Did We Learn?

Where Do You Find Rocks and Soil?

Think about the world around you. Where can you see rocks? How big are they? Are they all the same? Where can you see soil? What color is it?

Find rocks around you.

❶ Choose a place. List the rocks or things made of rock there.

❷ Now draw a rock picture. Include soil in your picture.

My Science Book

What Are Rocks Like?

Did all the rock and soil pictures look the same? Probably not. The world has many different kinds of rocks.

Shale

Marble

Sandstone

Granite

Basalt

Classify rocks.

❶ Use your hand lens to help sort the rocks into groups.

❷ Make a chart that shows how each group is different from the other groups.

One reason why rocks are different from each other is that they are made of different minerals.

But what is a mineral? It's matter that was never living and is found in the earth.

Amethyst

What's in a Rock?

All rocks are made up of two or more minerals. The many colors you see in rocks are all the different minerals. What other differences could the minerals cause?

Granite

Be a rock scientist.

❶ Choose a rock. Try to scratch it with your fingernail. Try the same thing with a nail. Chart your results.

❷ Test the other rocks and chart them, too.

Are some rocks harder than other rocks?

How Do Rocks Form?

All the different rocks you see form in the earth's outer layer, called the crust. New rocks form all the time, but they don't all form the same way.

Dirt, sand, clay, and broken rock get pressed together. They harden into sedimentary rock. 2

Any kind of rock buried many miles down can melt. Then it can become magma again. 4

1 There's hot liquid rock called magma under the crust. When magma flows out of a volcano, it's called lava. Lava cools into igenous rock. Volcanoes erupt all over the world. Popocatéptl, (po-po-ka-tay-páy-tull), a volcano in Mexico, erupted in 1996.

3 Heat and the weight of rock above can change rock buried deep in the crust into metamorphic rock.

Some rock changes take millions of years. A rock you pick up today might once have been buried deep in the earth.

My Science Book

How Do Rocks Form in Water?

Rocks form in layers within the earth's crust. Some of these layers are under oceans and lakes. How do rocks form in big bodies of water?

Shake a lake.

❶ Put equal amounts of sand, gravel, and dirt in the bottle. Fill with water.

❷ Put the cap on tightly and shake the bottle.

❸ Wait one hour.

Compare what's in your bottle to the picture on the right. How are they alike?

Sand, dirt, clay, and small bits of rock and shells settle to the bottoms of oceans and lakes. Over very long periods of time, under tons of water and mud, the bottom layers change into rock.

How Do Mountains Form?

Mountains are made of layers of rock. They are part of the earth's crust. How are mountains formed?

Make a mountain.

❶ Flatten three long pieces of clay, and lay them on top of one another. Press down a little to make them stick together. These are your layers of rock.

❷ Push both ends of your pile toward the middle. What happens?

How are your clay layers like the cliff above?

Mountains form when the earth's rock layers bend upward. You can't see or feel this happening because it usually takes millions of years for the layers to move.

What Happens When the Earth Shakes?

The earth's rock crust is always moving, up, down, and sideways. Sometimes this movement forms mountains or new rocks. But what happens when layers of the crust rub against each other?

Be an earth shaker.

❶ Put the two sides of the pan together and line it with foil.

❷ Fill the bottom of the pan with moist dirt. Now build a neighborhood.

❸ Quickly rub the two pieces of the pan against each other. What happens?

How Do Wind, Water, and Ice Change Rocks?

When the earth's crust moves, it changes rocks. Other forces change rocks, too.

Wind carries tiny grains of sand when it blows. The sand rubs against rocks and wears them down bit by bit. This is called erosion.

What do you think happened to these rocks?

When water and ice rub against rock, they wear away its surface. This is erosion, too.

What Is Soil?

The bits of rock that wear away become soil. In most places, soil is at the very top of the earth's rock layers.

What can you say about these soils?

Loam

Clay

Be a soil inspector.

❶ What does each soil sample feel like? How do the soil colors compare?

❷ Look at each soil sample with the lens and draw what you see.

Besides rock bits, what other matter did you see in the soils? Are there minerals in soil?

What do you think makes soil different in different places?

Sand

Why Is Soil Important?

Soil is found in many places. It's the thinnest of all the earth's layers. What are some ways people use soil? How do plants use soil?

Plant a field.

❶ Fill half the pan with loam and the other half with sand. Scatter grass seed evenly.

❷ Put your field in the sun, and water it every day. Write what happens after a few days and after one week.

Which one of your test soils is better for the grass? Why? Which soil would you want to use if you were a farmer?

What Does Erosion Do to Soil?

You saw that water and wind can erode rocks. What can water and wind do to soil?

Make an Erosion Model

❶ Mix the loam and sand in the pan to make a hill.

❷ Hold the paper behind the hill. Blow gently at the hill through the straw. What happens?

❸ Shake a wet sponge over the hill. Then pour some water on the hill. Now what happens?

How Can People Protect Soil?

Plants need good soil to grow. Erosion takes soil and minerals away from the land. Then the soil often moves into rivers and out to sea. What can keep soil from blowing and flowing away?

Save the soil.

❶ Shape your sand and loam into a hill again.

❷ Find some ways to keep wind and water from taking the soil away.

❸ Test your ideas. Shake the wet sponge over the hill. Then pour a little water on the hill. What happens each time?

What Can You Learn from Rocks and Soil?

Scientists learn about the earth by studying rocks and soil. What clues do you think earth scientists look for?

Make your mark.

❶ Flatten a piece of clay. Press your secret object into it. Remove the object carefully.

❷ Pour wet plaster into the print. Wait 20 minutes. Remove the plaster. What do you see?

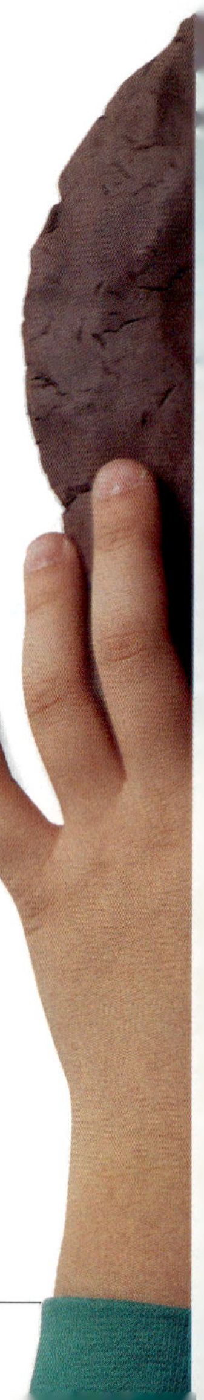

My Science Book

▶ Camarasaurus fossil

Camarasaurus

Long ago, dead plants and animals fell into mud. Other animals left their tracks in mud. Then more mud covered the bodies and tracks. Over millions of years, the mud hardened into rock. The marks left by the animals and plants became fossils.

What's the Rocks and Soil Story Where You Live?

What are the rocks and soil like where you live? How did wind, water, ice, or people change them?

Make an earth mobile.

1. Decide what pictures you want to draw to tell the story of the rocks and soil where you live.

2. Draw each picture and write about it. Put the pictures in order and tape them all together to make a circle.

3. Tape string to the inside so your circle will turn when you hang it up.

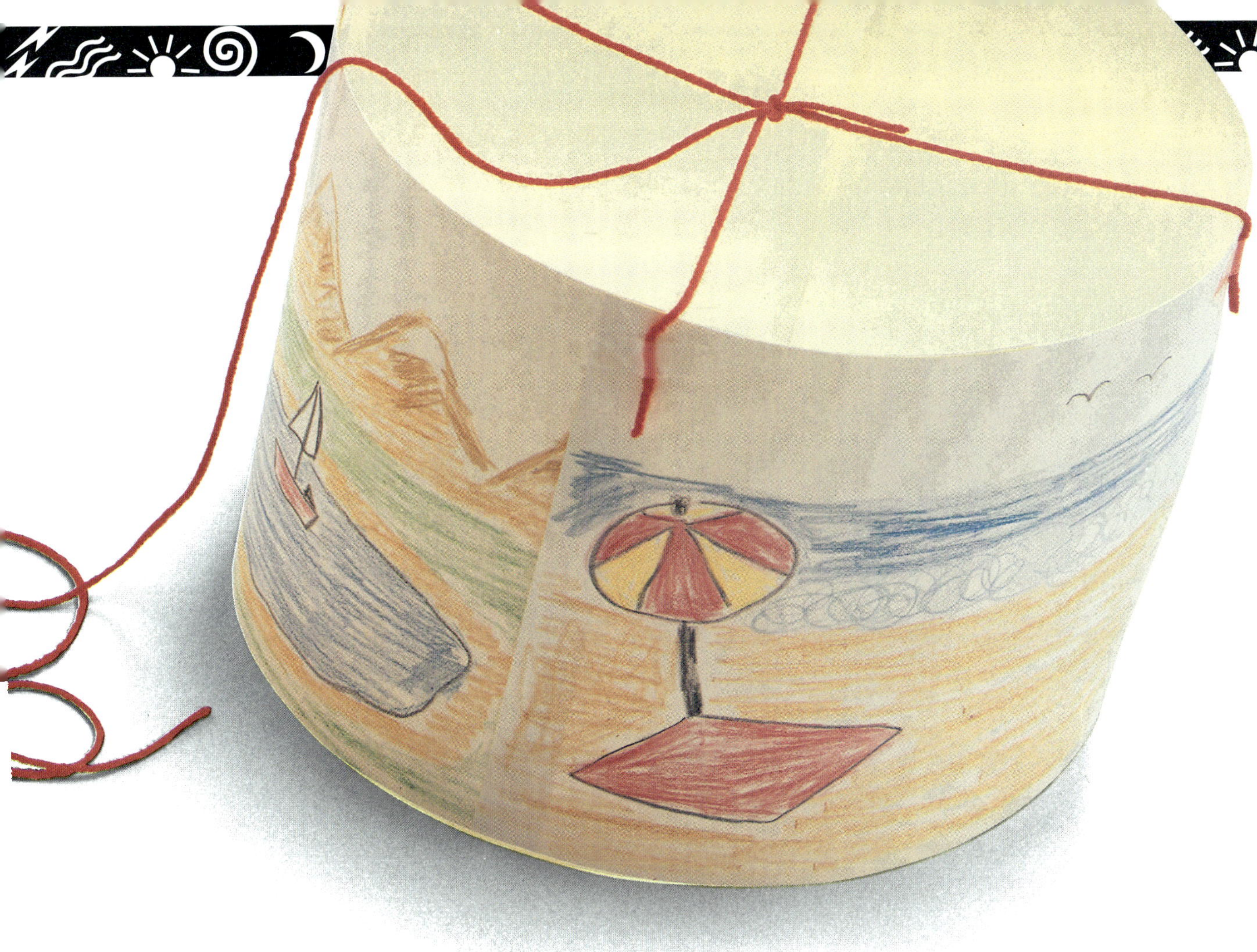

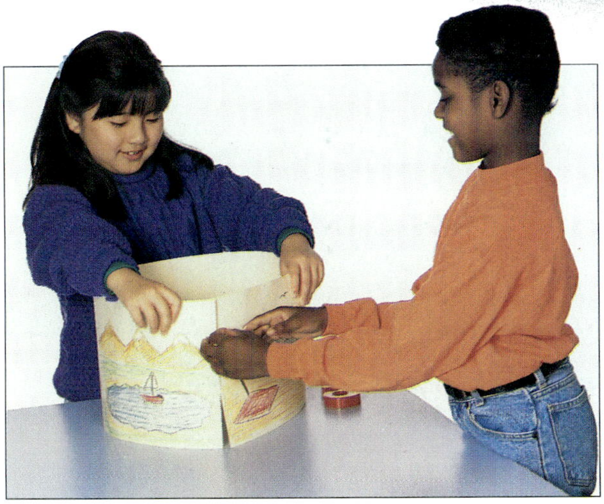

Over and over, the earth keeps on changing. Rocks break down and form new soil. Soil and rock bits settle into layers and form new rocks. Like a great big circle the story never ends.

Crust: The outer layer of the earth is called the crust. The crust is a layer of rock and soil about 25 miles deep. It includes all mountains, hills, and valleys.

Erosion: Erosion is the blowing or washing away of rock or soil on the earth's surface. People plant gardens and farms in special ways to prevent erosion.

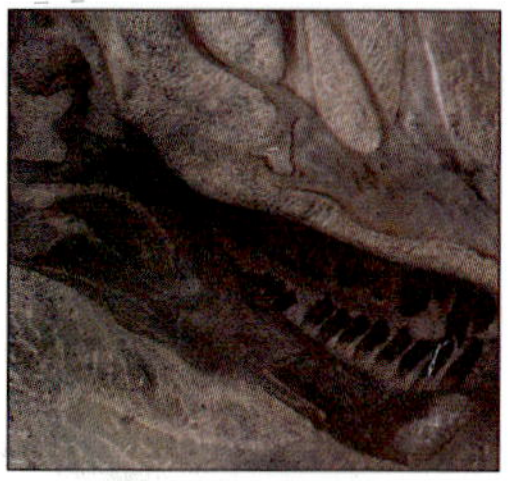

Fossil: A fossil is a bone or shell turned to rock or a mark made in mud by a plant or animal and turned into rock over thousands of years.

Igneous rock: Igneous rock forms when melted rock cools inside the crust or when lava flows out of a volcano onto the earth's surface.

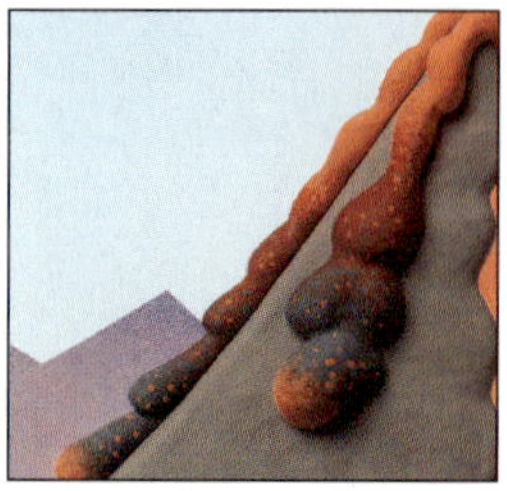

Lava: Lava is hot liquid rock that has reached the surface of the earth. Flowing lava looks like a river of fire.

Metamorphic rock: Heat and pressure deep inside the earth's crust can change one kind of rock into another kind. The new kind of rock is called metamorphic rock.